Walter Prentis

Notes on the Birds of Rainham

Including the District Between Chatham and Sittingbourne

Walter Prentis

Notes on the Birds of Rainham
Including the District Between Chatham and Sittingbourne

ISBN/EAN: 9783744679039

Printed in Europe, USA, Canada, Australia, Japan

Cover: Foto ©Thomas Meinert / pixelio.de

More available books at **www.hansebooks.com**

THE BIRDS OF RAINHAM.

NOTES

ON THE

BIRDS OF RAINHAM

INCLUDING THE

District between Chatham and Sittingbourne

BY

WALTER PRENTIS

LONDON

GURNEY & JACKSON, 1, PATERNOSTER ROW

(SUCCESSORS TO MR. VAN VOORST)

MDCCCXCIV.

LONDON :

PRINTED BY WOODFALL AND KINDER,

70 TO 76, LONG ACRE, W.C.

INTRODUCTION.

Rainham is situate half-way on the high road between London and Dover, 4 miles from Chatham, and 6 miles from Sittingbourne ; it is said that this road was at one time the Watling Street of the Romans ; an extremely early British gold coin was picked up here a few years ago ; it is also said that Cæsar, after defeating the Britons in East Kent pursued them through Rainham across the Medway and Thames into Essex.

Rainham lies in the North-Eastern Parliamentary division of Kent. The Church stands on rising ground on the south side of the high road in the centre of the parish, and is an attractive edifice, more especially from the railway ; it is said to have been built in the 13th century ; taking a survey from the top of its handsome beacon-tower a fine prospect over a limited extent of country is obtained ; looking towards the east, standing on high ground the first object which strikes attention is the parish church of Upchurch with its reduced spiral steeple ; beyond you will have the pleasure of seeing in the distance the

6 INTRODUCTION.

isle of Sheppey, Queenborough, Sheerness and the Nore.

Looking north-east you will behold the mouth of the Medway, the isle of Grain, and on a clear day Southend in Essex. Turning to the north you will observe more diversified scenery, the busy haunts of man directly in front of you in the London Chatham and Dover railway, a mile of modern brick-built houses, the railway-station leading the way, a little further on you will perceive some ploughed fields, brick fields, a cement factory, orchards and marsh ; lying between two creeks in the distance is a high bank of fine sand which must have been thrown up by the water at the last formation of the earth's surface ; again extending your view, is the river Medway with its numerous craft, the land further beyond is the hundred of Hoo.

Again looking in the direction of north-west you will see in the distance the beacon-tower of Gillingham Church ; beyond, on the opposite side of the Medway, are the hills of Upnor, adjacent to the city of Rochester and the dock-yard of Chatham ; turning round to the west and south there is no view beyond the woods, and the orchards three parts encircle the foot of the church.

The soil of Rainham is a good loam resting upon chalk, more or less deep where the chalk does not approach the surface, more or less thin near the woods where the soil is flinty. Brick-making has been carried on for some few years, there are three wharves at the terminus of two creeks which extend one mile from the Medway. Clay-digging upon the salt marshes is another trade for the purpose of making cement; barges sail to and from London and elsewhere loaded with bricks, coal, clay, hay, straw, and most other kinds of produce and manure; the latter is a great benefit to the parish in every way rendering the soil more productive. Cultivation is going on as well as it can be done, almost everything that is either sown or planted grows as a rule to an average perfection.

Rainham possesses 2000 acres of land containing a population of 3000 inhabitants, swarms with children, 700 of whom attend school; this must be considered a thriving parish and cannot be said to stand last upon the list towards national prosperity. Were the people evenly distributed they would occupy every acre, fortunately for the naturalist they prefer not the woods on one side of the parish or the marshes on the other; they flock for the most part into the principal village on the high road in the centre of the parish, where the shops, school, church, and various other places of resort are situate; employment is provided for all who are disposed to work, either in field or factory, high, low, rich and poor alike.

I now come to the subject of my text—Notes on the Birds of Rainham including the district between Chatham and Sittingbourne.

That extremely wet summer of 1860 is my excuse, or rather my pleasure, for making birds a study. Generally at home on my farm in the country my opportunities have not been slight for bird observation, at the same time being fond of my gun with a

predilection for natural history, I have followed the
pursuit on my own and neighbouring farms including
the district from the year 1860 to the year 1894.

I do not know but that my little book may con-
tribute something towards a future history of the
birds of Kent.

RAINHAM, *June,* 1894

THE
BIRDS OF RAINHAM.

———•———

SEA EAGLE.

Haliaëtus albicilla (Linnæus).

A pair of Sea Eagles paid my district a visit in the
month of November, 1879, one of them while sitting
upon an oak tree over a furze bank frequented by
rabbits, soon fell a victim to a sportsman's gun, a fine
large Eagle in mottled plumage and with a whitish
tail :—Another eagle was soon afterwards seen flying
about 150 yards high over the district ; throughout
this very severe winter it was seen to frequent
manure heaps, killed a rook caught in a trap, and fed
chiefly upon the marsh hares, carrying them over the
marsh walls where generally a good look out was
kept ; however, in course of time, a shepherd with a
gun approached, when too near to be pleasant his gun
happened to miss fire, the Eagle flew away and was
no more seen.

OSPREY.

Pandion haliaëtus (Linn.).

I had the pleasure of meeting with an Osprey
sitting upon a post on the salting in the parish of
Upchurch, adjoining Rainham ; it had been feeding
upon a flounder and when shot at by myself and
friends it escaped, taking a direct course up the river ;
an Osprey was shot three days afterwards, according
to the " Zoologist," at Uxbridge in Middlesex.

An Osprey was shot here by Mr. George Power
about the year 1878 in September—the month when
the Grey Mullets are plentiful ; it was seen sitting on
the edge of a narrow piece of salting beside the marsh
wall, over which it was shot ; at the same time being
mobbed by a flock of Peewits.

PEREGRINE FALCON.

Falco peregrinus, J. F. Gmelin.

There is seldom a year passes that one or more
Peregrines are not seen flying over the Medway, they
are chiefly young birds, a few adults are now and then
obtained ; according to my notes one of the latter to
five of the former ; it is a very interesting sight to
witness a Peregrine Falcon in pursuit of a flock of
Starlings, dashing amongst them in the air until one
is singled out for prey.

HOBBY.

Falco subbuteo, Linn.

A pair of Hobbies in the year 1860 took possession of an old Magpie's nest on the top of a row of Elm trees standing in one of our lanes ; they were supposed to be Sparrow-hawks, but that could not be owing to their long wings and powerful flight, they seem to have brought off their young in safety ; I did not happen to see them at the time, but we were visited by Hobbies for two years afterwards.

In the month of June, 1864 a male Hobby was shot in a cherry orchard this side of Sittingbourne by the bird-scarer, and the female was seen.

MERLIN.

Falco æsalon, Gmel.

A few years ago our woods were frequented by Merlins, they always came to roost in the winter time at a quarter to four o'clock, somewhat later when the days got longer. I once observed five altogether sitting upon a small oak-tree standing by itself in a low coppice ; a pair continued to frequent the wood for some years afterwards, coming in to roost at the usual time, the female always before settling down to roost for the night took her station on the top of an elevated perch where a good look out was kept ; the male at the same time keeping well out of sight amongst the thick high coppice.   When the month of

April came round and just before their departure, I observed both of them sitting upon the same tree, one on each side of it.

KESTREL.

Falco tinnunculus, Linn.

The Kestrel I observe occasionally, they appear to be not so common as they should be, I believe the reason is, that owing to our "pop-gunners" in the winter time, the poor Kestrel flying rather high and always being in view, gets shot; they seem to take possession of an old Magpie's nest if possible, and to breed near the outside of a wood.

LESSER KESTREL.

Falco cenchris, Naumann.

The reader may wonder how I became acquainted with such a bird : my answer is, I saw one in the Museum at Dover. What I am writing is about another which I observed, exactly like it. The latter end of March, 1882, a pair of Linnets appeared in my garden as usual ; on the 3rd of April I saw a lump of Linnet's feathers on my lawn, such as an inverted tea-cup would cover, which must have been done by some hawk. Having never seen the like before I was much puzzled, however on the 10th of April a hawk-like chatter came issuing from a cluster of elm-trees standing in the field beside my garden, when a beautiful little fellow with sharp-pointed wings flew nearly over my head, circling and wheeling some thirty-five yards high ; the sun shining brightly at the same time, I had a good view of its back, which was red, the bird stayed here for nearly a month and after being shot at, becoming wilder and wilder each day.

SPARROW-HAWK.

Accipiter nisus (Linn.).

This, our most common hawk, makes its appearance in the autumn, staying with us, if not shot, throughout the winter ; where the small birds congregate on the stubbles, there is often a female Sparrow-hawk in attendance, the little red males lying wait in the thickets for the chaffinches or any passing prey.

I once saw a female Sparrow-hawk flying away from a clump of larch fir-trees, carrying material to build her nest, with quite a little faggot in her claws.

An immature Sparrow-hawk was picked up dead by a neighbour beside her parlour-window, having flown at a canary in a cage.

BUZZARD.

Buteo vulgaris, Leach.

The year 1870, the time of the French and German war, was famous for being our Buzzard year ; according to Mr. Charles Gordon of the Dover

Museum 25 Buzzards were obtained in East Kent
alone ; I have heard of others being obtained in West
Kent, a pair was shot in my district.

The same year I happened to hear a great clamour
overhead, on looking up I saw about a dozen Jackdaws mobbing a Buzzard, the day was very fine and
warm ; they pursued their course upwards higher
and higher, at last the Jackdaws having succeeded in
driving the Buzzard away tumbled down one after
the other into their natural element, contented.

ROUGH-LEGGED BUZZARD.

Buteo lagopus (Gmel.).

A Rough-legged Buzzard was shot December 31st
1879, by a woodman in pursuit of Pigeons, on a wet
misty day ; it flew from one tree into another where
it was killed. I observed another about the same
time on our salt marshes.

HONEY-BUZZARD.

Pernis apivorus (Linn.).

A pair of Honey-Buzzards were shot here September 1881, both in immature plumage ; one a
deep chocolate colour, the other a spotted variety
beautifully variegated with white.

MARSH-HARRIER.

Circus æruginosus (Linn.).

A Marsh-Harrier was shot and obtained in our woods in June 1867, of a deep chocolate colour, head included, just beginning its moult.

HEN-HARRIER.

Circus cyaneus (Linn.).

Females are not uncommon in the autumn time of the year frequenting our woods, where they are sometimes met with and shot ; the old grey males are seldom or never seen.

I once, when driving in the month of August, had the pleasure of seeing a pair, an old grey male and brown female, circling round the centre of a field of barley adjoining our main road.

ASH-COLOURED OR MONTAGU'S HARRIER.

Circus cineraceus (Montagu).

The Ash-Coloured Harriers, when they arrive, always come to us in the spring of the year, unlike the Hen-Harriers, which always come in the autumn ; they at once take up their quarters in the woods, flying over the fields and low coppices, till they are shot, they seem to vary much in plumage ; the name Ash-Coloured applies to the old male with black bar across its wings ; 6 varieties have been obtained in my district.

No. 1. Old male with a black bar across wings.

No. 2. Old female dressed in shades of brown.

No. 3. Immature male, ash coloured with some light brown feathers intermixed.

No. 4. A dark coloured variety, chocolate, almost black.

No. 5. Dark chocolate mixed with brown.

No. 6. Light brown on the back, on the breast white slightly tinged with red.

Perhaps it may be of interest to note the date of each occurrence.

No. 1. June 8, 1866.	No. 4. May 18, 1867.
No. 2. May 17, 1869.	No. 5. July , 1870.
No. 3. May 15, 1866.	No. 6. May , 1888.

LONG-EARED OWL.

Asio otus (Linn.).

Not so common with us as the next to be described, they are shot in our woods occasionally, and on one occasion when hunting with beagles several flew from a leafy oak. I have never heard of their breeding in my district, we have no fir plantations.

SHORT-EARED OWL.

Asio accipitrinus (Pallas).

Not uncommon, comes in the autumn, visits our salt-marshes where they are shot nearly every year.

When partridge shooting I have met with them in our turnip fields ; on one occasion a pair nested and succeeded in hatching their young on an island marsh which had been lying idle throughout the winter and spring.

BARN OWL.

Strix flammea, Linn.

This most useful bird, I am sorry to say, is a victim to persecution by every pop-gunner in the parish on coming, as it sometimes does. Its life after visiting our stacks once or twice is sure to be sacrificed. I have known a pair to breed more than once in our church roof, but this was a few years ago.

GREAT GREY SHRIKE.

Lanius excubitor, Linn.

The Great Grey Shrike has been observed and shot in my district at least three times.

I had the pleasure on one occasion of observing a Shrike on the top of a faggot stack in a wood ; it was a brown bird, not unlike a Thrush, but when it flew off it settled upon the tops of the high spray and litters round an adjoining wood where I followed it ; I had no difficulty in identifying the bird, especially when it hovered like a Kestrel over a grassy lea. Could it have been a young bird? if so it was very wild. Yarrell does not speak of such a variety, neither do other ornithologists.

C

RED-BACKED SHRIKE.

Lanius collurio, Linn.

Comes to us the latter end of April or the beginning of May, is not uncommon and breeds if not disturbed, as a rule in thick thorn bushes on the borders
of pastures where beetles and other insects abound,
but this is not always constant; I once found their
nest in a roadside hedge bordering a ploughed field ;
pastures were close at hand.

WOODCHAT SHRIKE.

Lanius rutilus, Latham.

One was shot May 7, 1868, not exactly in my
district, in an orchard ; in its crop, as I was
informed, was a bee, some caterpillars, and two pieces
of grit ; it is a light coloured specimen, probably
immature.

SPOTTED FLYCATCHER.

Muscicapa grisola, Linn.

The Flycatcher comes to our homesteads and
orchards when the apple trees are in blossom, stays
with us throughout the summer doing all the service
it can, leaves sometime in September ; I, for one,
wishing it a safe passage all the way to Africa and
back again.

Pied Flycatcher.

Muscicapa atricapilla, Linn.

I have never once met with the Pied Flycatcher. One of a pair was shot in a wood-lane, by a small pond, May 1st, 1871.

Missel Thrush.

Turdus viscivorus, Linn.

A few mild days in February the Missel-Thrush begins to enliven us with its song, being the first of the Thrushes to remind us of the time of the year ; builds the latter end of March, making its nest and laying its eggs in either an apple or pear tree in our orchards, but owing to the bare foliage is seldom allowed to rear its young ; it never attempts to make a second nest near the same place.

The loud wild note of the " Storm Cock " is most engaging, as a rule indicative of rough stormy weather.

Song Thrush.

Turdus musicus, Linn.

The Song thrush begins to warble forth its charming song the beginning of February, and enlivens us at bright intervals throughout the month ; when spring time comes round the Thrush sings its wild, loud, melodious carol on an elevated perch from morning till night. They suffer much in severe weather which carries many off ; no bird would be more missed than the Thrush ; I have always, I am happy to say, one which has withstood the chills of winter left in my garden to sing through the spring and summer, and to share with me the fruits in my garden.

FIELDFARE.

Turdus pilaris, Linn.

Some winters uncommonly plentiful, others very scarce, this appears to depend on their immigration, north or south ; I have known them some seasons to arrive in continuous flights, one after another, in the same direction all the day long, the last flight in the dusk of the evening coming in contact with the telegraph wires on our railway, where five were once picked up at the same time ; they appear on this occasion to have come by way of the Isle of Sheppey, taking a line parallel with the creeks towards our lower orchards, and after topping them flying direct for the woods, where they turned towards the west into the country. Should they take a northern passage across the sea, we see but few throughout the winter. A cream coloured variety was shot December, 1878.

REDWING.

Turdus iliacus, Linn.

Comes in the autumn, the latter end of October, in small numbers, feeds on our pasture land and in our orchards ; when the snow lies on the ground they feed upon the hawthorn berries and on the manure heaps. Should the winter be severe they suffer equally with the Song Thrush.

RING-OUZEL.

Turdus torquatus, Linn.

The Ring-Ouzel passes through Rainham on its passage north in spring, south in the autumn, sometimes, when food is to be had, staying a few days with us. Always wild, choosing for its perch the tops of trees.

BLACKBIRD.

Turdus merula, Linn.

The Blackbird frequents our gardens all the year round, at no time does its jet black plumage shine with more lustre than when in mid-winter the snow is lying on the ground, associating with the sparrows and feeding where the poultry are fed.

HEDGE SPARROW.

Accentor modularis (Linn.).

The Hedge Sparrow is another of our garden birds almost always to be seen on looking out of the window, picking up something on grass plots and gravel walks.

ROBIN.

Erithacus rubecula (Linn.).

The familiar Robin makes itself at home everywhere.

NIGHTINGALE.

Daulias luscinia (Linn.).

Everybody knows the coming of the Nightingale, soon as springtime comes we are waiting for the great event of the year to hear our woods and hedges rattle with its glorious song; the middle of April is the time, nor does it lose a day or night after coming, if the weather is at all favourable, to pour forth at once its charming melody.

What would a country life be worth in the south of England without the far-famed Nightingale?

REDSTART.

Ruticilla phœnicurus (Linn.).

The Redstart once was common in our orchards, coming the second or third week in April, now only to be heard and seldom seen on the outskirts of the woods, where it breeds in the old boundary stumps.

BLACK REDSTART.

Ruticilla titys (Scopoli).

When driving, October 1st, 1865, I observed a

Black Redstart with white on its wings, fly along a wire fence.

An immature Black Redstart was caught by a bird-catcher in one of our brickfields, Nov. 1887.

STONECHAT.

Saxicola rubicola (Linn.).

The Stonechat is more common with us in winter than in summer. I have never heard of its breeding in my district, but it may possibly do so in some snug corner or other. We have no furze banks or rough commons, and it does not appear to approve of clean pastures and cultivated fields.

WHINCHAT.

Saxicola rubetra (Linn.).

Comes in April, probably breeds in my district more often than the Stonechat; a pair frequented a narrow coppice on my farm in 1886. On mowing clover and making hay in the field adjoining, a nest was discovered containing four blue eggs, built on the ground, after the manner of the Skylark, the scythe cut clean over it; the hay-makers were puzzled, never having seen blue eggs in such a place before, at least fifty yards from the hedge.

WHEATEAR.

Saxicolo œnanthe (Linn.).

The Wheatear, coming in March, is the first spring
arrival which attracts our attention ; more often seen
on the marsh than on the plough land. I have
noticed some few in summer time flitting along our
marsh walls showing their conspicuous white tail
coverts, probably breeding in the blocks of stone used
for supporting the embankment.

GRASSHOPPER-WARBLER.

Acrocephalus nævius (Boddaert).

Some six years ago before the cold wet summers set
in, our woods were alive with ' cricket birds,' their note
could be heard on a summer evening in all directions ;
in consequence of the cold and wet lasting year after
year they have become scarce, and last year not a
' cricket bird' was to be heard, they have, in fact
totally disappeared.

A singular variety was obtained June 5th, 1869,
back greenish brown, with darker markings, breast
greenish yellow without spots, a male, shot singing.

SEDGE-WARBLER.

Acrocephalus schœnobœnus (Linn.).

Comes in April, heard and seen in our marshes,
frequents ditches which are overrun with brambles or

any rough herbage ; this is generally the case where the ditches divide the marshes from the plough land.

REED-WARBLER.

Acrocephalus streperus (Vieillot).

The Reed-warbler, as its name implies, is never met with as a rule beyond the reach of a spot where reeds grow ; they seem to know the time when the reeds afford them shelter and protection in May. I have a specimen of the Marsh-Warbler—if the distinction is a broader bill than that of the Reed-warbler—which was shot in a garden beside a reed bed at Milton, May 1866.

BLACKCAP-WARBLER.

Sylvia atricapilla (Linn.).

The merry Blackcap comes in spring rather earlier than the Nightingale, leading the way and preparing us for the advent of its superior rival ; at the same time the note of the Blackcap is one among the many Warblers that come afterwards, and stands next to the Nightingale. He frequents our gardens, sings close to our doors directly on coming, making us at once acquainted with his presence.

GARDEN WARBLER.

Sylvia salicaria (Linn.).

The Garden Warbler was frequent in our gardens before the setting in of the cold wet summers, since

which time they have been just as scarce as they
were plentiful. I now hear only here and there one
in our coppices and woods.

WHITETHROAT.

Sylvia rufa (Bodd.).

The Whitethroat is, perhaps, the most common of
all our summer warblers, comes as a rule the third
week in April, frequents the roadside hedges, con-
structs a thin flimsy nest made of the cleavers or
goosegrass a foot from the ground, utters a merry,
cheerful note, is provincially called jolly Whitethroat.

LESSER WHITETHROAT.

Sylvia curruca (Linn.).

Not very numerous as a species. A pair almost
always come and sing near my dwelling when the
apple trees are in bloom, sometimes breed in the
garden shrubs.

WOOD-WARBLER.

Phylloscopus sibilatrix (Bechstein).

The Wood-Warbler is seen on passage ; our woods,
being cut down every few years for the sake of the
hop poles, do not appear to be adapted to their
nature.

WILLOW-WARBLER.

Phylloscopus trochilus (Linn.).

The Willow-Warbler ere the setting in of our wet summers, was very numerous, as a species no bird excepting the Grasshopper-Warbler seems to have suffered more from that cause; still we have a few left, which can be heard in our woods. Surely it must take several years of warm weather to recruit the ranks of our summer Warblers.

CHIFFCHAFF.

Phylloscopus collybita (Vieill.).

The Chiffchaff comes into our woods early, sometimes before the buds have burst into leaf, where its wellknown chiffchaff note wakes us up, being the first messenger of spring's glad tidings for the future, should the summer prove acceptable to vegetation and to bird life.

GOLDEN-CRESTED WREN.

Regulus cristatus, Koch.

I have not seen the Golden-Crested Wren for several years, what has become of them I do not know, unless they have been affected by the cold and wet summers. On one occasion a pair built their nest in an Irish Yew tree in my garden.

FIRE-CRESTED WREN.

Regulus ignicapillus (Brehm.).

My first acquaintance with the Fire-Crested Wren
was when I first thought of making a small col-
lection of birds.   In that wet year 1860, anxious to
obtain a Goldcrest, I thought I saw one in an apple
tree beside my house ; on picking the bird up, to my
delight, it was a female Fire-Crest.   Date of occurrence, December, 1860.   Since, I have once met
with a pair in the month of October on some larch
firs.

GREAT TIT.

Parus major, Linn.

I observe these sprightly little birds in small
parties when the woods have shed their leaves. When
the foliage is thick they are scarce, we have no old
timber trees for them to nest in. They do not breed in
our orchards.

BLUE TIT.

Parus cæruleus, Linn.

In winter the Blue Tit does not appear to be so
numerous as the Great Tit.   In summer they are at
home near our dwellings, and are with the rest of the
Tits examples of industry, never idle from morning till

night, searching the twigs in every direction for insect food, which they entirely depend upon for subsistence.

COLE-TIT.

Parus ater, Linn.

I meet with the Cole Tit but rarely, perhaps one in the course of a year; they are always the continental variety. There are no fir plantations in my district to harbour the common species.

MARSH TIT.

Parus palustris, Linn.

I scarcely ever meet with the Marsh Tit in the marsh. They breed in the low stubs in our woods, and in winter time when the snow is on the ground, I observe them feeding among the horse-droppings in the roads by the woods.

LONG-TAILED TIT.

Acredula caudata (Linn.).

The Long-tailed Tits, like some of the warblers, have become very scarce. I seldom meet with them now. Before the wet and cold summers set in they were plentiful, building their beautiful domed nests in our woods.

BEARDED TIT.

Panurus biarmicus (Linn.).

The Bearded Tit has not, to my knowledge,
occurred in my district.   Three were shot in a reed
bed on the banks of the Medway in the winter of
1865, near Maidstone.

WAXWING.

Ampelis garrulus, Linn.

The Waxwing is extremely rare in this district,
only two to my knowledge have been obtained ; one
was shot from a pair in our woods in 1867, another
obtained in an orchard the same year.   A small
flight was seen in an orchard at Milton, taking a contrary direction from here where one or two others
were obtained.

PIED WAGTAIL.

Motacilla lugubris, Temminck.

The neat chaste Wagtails keep us company all the
year round, being most numerous in the autumn.
They visit our sheep-folds in the winter, running
among the sheep while feeding on the turnips.   In
summer we see them following the plough, picking up
the wire-worms in the furrows, going and coming
with their bills quite full of them for several hours
together, and continuing their good offices till the
team leaves off work.

Grey Wagtail.

Motacilla sulphurea, Bechs.

The Grey Wagtail does not to my knowledge breed in this district. I now and then meet with one or two in the winter time running along a fresh water rivulet or beside a marsh ditch.

Yellow Wagtail.

Motacilla Raii (Bonaparte).

The Yellow Wagtail comes on the 10th of April, sooner or later according to the weather; the first comers are always the most brilliant in plumage. After a day or two's stay they go north followed by others which are all young birds, and such we are obliged to make ourselves contented with ; they breed as a rule in our pea fields, not often in the clover.

Tree-Pipit.

Anthus trivialis (Linn.).

The Tree-Pipit is one of our summer arrivals ; comes in mid-April, frequents trees in our low coppices, whence circling upwards and returning to the same perch, sings in the air most sweet and delightful notes. When making hay near the woods the Tree-Pipit flies from heap to heap keeping the haymakers company, singing all the while its pleasing song.

MEADOW-PIPIT.

Anthus pratensis (Linn.).

The Meadow-Pipit is not very common in my district, more so in winter than in summer. We are without much rough herbage in spots on the meadows where they delight to frequent.

RED-THROATED PIPIT.

Anthus cervinus (Pall.).

I fell in with the Red-throated Pipit one fine sunshiny day at the beginning of the month of April, 1880, flying up and down, singing and feeding along the ploughed furrows behind my plough while turning over a two years grassy lea. I was attracted by the bird being alone, and returned with my gun and shot it. On picking it up I thought it could be nothing more that a bright example of the Meadow-pipit. Fortunately, I sent it to Mr. Gordon of the Dover museum to be preserved.

ROCK PIPIT.

Anthus obscurus (Lath.).

The Rock-Pipit is common in winter along the shores of our creeks, picking up something on the edge of the water as the tide recedes from the shore; does not occur in summer time.

I once had the pleasure of meeting with the Vinous Pipet (*Anthus rupestris*) one rough March, the wind blowing for several days from the east the marsh was full of rock Pipits ; I observed a strange looking variety with a reddish-brown breast, sitting upon a barway ; its note, on being disturbed, was very different from the Rock-Pipit's.

RICHARD'S PIPIT.

Anthus Richardi, Vieill.

I have only upon one occasion had the pleasure of meeting with Richard's Pipit. I observed a long-tailed lark sitting on a gate in an island marsh when the tide had covered the saltings, it flew from off the gate to underneath a lump of deal plank which had been used for repairing the marsh wall, from thence into a cabbage garden and out of sight beside a pen of sheep. Surely, Richard's Pipit must be a very creeping bird.

SKYLARK.

Alauda arvensis, Linn.

The Skylark is everywhere plentiful. The first fine day or two in February he begins to sing, never failing to do so all the year except in rough weather and moulting time ; just before a thick fall of snow they migrate, but return again soon after a thaw.

D

WOODLARK.

Alauda arborea, Linn.

The only time I have met with the Woodlark was when the labourers were throwing out the heaps of manure in winter, when the snow lay thick on the ground and the woods were covered with ice ; its note attracted my attention ; others have occurred at the same season of the year.

SNOW-BUNTING.

Plectrophanes nivalis (Linn.).

The Snow-Bunting come in flocks about November and December ; they appear to be young birds generally, probably some females amongst them ; the old males seem to come more singly, they mix with the skylarks and are known by their white plumage.

CORN-BUNTING.

Emberiza miliaria, Linn.

The Corn-Bunting is spread over all the district, not very numerous as a species, generally to be met with on search being made, does not associate as a rule with other birds, breeds in the grass when the meadows are laid in for hay ; also in pea-fields, singing its well-known note on the top of a bush or fence.

Black-headed Bunting.

Emberiza melanocephala (Scop.).

This Bunting is almost if not entirely peculiar to
the reeds, rushes and sedge in the marsh, where they
are always to be found ; they build their nests in low
bushes. I once saw one in some dead herbage thrown
up by the tide.

Yellow Bunting.

Emberiza citrinella, Linn.

The Yellow Bunting congregates with the Finches
on the stubble fields in winter, and not till the hedges
are well secured in leaf does the "Yellow-ammer"
think of building its nest in a low bush or hedge-bank,
oftentimes beside a public road. It flies from twig to
twig, or often upon the telegraph wires in front of you
as you walk, ride, or drive along the road.

Cirl Bunting.

Emberiza cirlus, Linn.

There is only one patch in my district where the
Cirl Bunting is to be found, about a mile of trees and
bushes bordering the marshes in the vicinity of two
old decoy ponds ; I once found a nest containing eggs
in an old dried up decoy pond, it was built on the

ground amongst the dead stems of the reeds. I have heard their note in the thick foliage on the top of trees ; they are sometimes caught by the bird-catchers in winter.

CHAFFINCH.

Fringilla cælebs, Linn.

Chaffinches having spent the autumn and the winter associated with other finches, feeding upon seeds on the stubble fields, and having paired, return in the spring with their merry note and shining plumage, building their beautifully constructed nests among the apple-blossoms in our orchards.

BRAMBLE-FINCH.

Fringilla montifringilla, Linn.

The Bramble-Finch is not very common in my district, not more than two or three being seen at the same time, generally mixed with the Chaffinches and other indigenous varieties on our stubbles, and known by their white tail coverts. I have on two occasions seen one from my window feeding with the sparrows ; they are fond of a newly-sown oat-field in the month of March, just before they take their departure to spend the summer in Norway.

HOUSE-SPARROW.

Passer domesticus (Linn.).

The bold hardy sparrows, no weather however severe affects them ; they swarm round our stacks and farm buildings in the winter, and when the summer comes they no doubt pick up insects for their young, redeeming in a small measure their offences. Corn is the food of the sparrow, they will rob the poultry if they can, no time of year are they so mischievous as when the corn is approaching maturity before harvest ; this is the sparrows' feast, they make the best use of it, they assemble together in flocks with their young, doing the farmer much damage almost without a cure ; white and cream-coloured varieties are often met with.

TREE-SPARROW.

Passer montanus (Linn.).

The Tree-Sparrow is common in the autumn and winter, arrives in October and November, keeps company with the finches on the stubble fields and round the corn stacks. I have observed them in small flocks by themselves on the marsh walls and beside the stable doors ; they do not appear to breed in my district, though a pair or two may possibly have done so.

GREENFINCH.

Coccothraustes chloris (Linn.).

The Greenfinch resorts to the stubble fields in autumn and winter, mixes freely with the Chaffinches and Yellow Buntings, Tree-Sparrows and Bramblings ; when the snow lies on the ground they visit the farm homesteads ; after the snow is gone join company again in the fields. They are rather bold birds, a horse-hair springe is sure to catch them, in summer building in rough hedges ; they are not so familiar as the Chaffinch and Yellow Bunting, but where the one is frequent the others are never far away.

HAWFINCH.

Coccothraustes vulgaris, Pall.

The Hawfinch is far from being an every day bird with us. On one occasion, in spring, I observed five busily feeding underneath some cherry trees in an orchard ; and upon another occasion a pair nested and reared their young in a wood beside a small pond ; single birds are sometimes met with.

GOLDFINCH.

Carduelis elegans, Stephens.

I do not always see a Goldfinch in the course of a year ; they seem to be getting more and more scarce.

A few teasles sometimes grow in our low coppices, where they are perchance met with.

SISKIN.

Carduelis spinus (Linn.).

Only upon two occasions have I met with the Siskin. Our chalk district does not possess any fresh-water streams to encourage the growth of the alder, the seeds of which I believe to be the food of the Siskin in winter.

AFRICAN WHYDAH FINCH.

Vidua.

A Whydah Finch was shot here. With the exception of one long tail feather being shot asunder it was in perfect plumage ; how it got into my district is difficult to say, it did not look like an escape.

REDPOLL.

Linota rufescens (Vieill.).

I have only met with a few solitary Redpolls feeding on rough herbage beside the roadside ; as with the Siskin, the district is not adapted to their requirements.

LINNET.

Linota cannabina (Linn.).

The Linnet is pretty generally distributed, fond of keeping their own company, are remarkably fond of turnip seed ; when it can be procured will live entirely upon it. They nest sometimes in low hedges and gardens, singing and cheering us with their sweet notes. Two white varieties have been obtained.

TWITE.

Linota flavirostris (Linn.).

The Twite comes to us in the autumn and winter, is always to be met with along our marsh walls, sometimes in hundreds, and feeds upon the seeds of the salt-water herbage.

BULLFINCH.

Pyrrhula Europæa, Vieill.

One of our most persecuted birds is the Bullfinch ; they leave the woods in winter for the plantations where their note is imitated, and they are shot, scarcely a single one escaping. Further back in the woods, out of reach of the plantations, when game shooting, a few may be seen.

CROSSBILL.

Loxia curvirostra, Linn.

Only upon one occasion in the space of more than thirty years have we been visited by Crossbills. On December 10th, 1868, a little family of about a dozen appeared in our nurseryman's garden, seen feeding on the berries of an Arbor Vitæ tree, in all shades of plumage, one red, three green, the rest various. On laying aside one of the females that was shot, contrary to the nature of things the body became quite dried up without turning in the least putrid.

STARLING.

Sturnus vulgaris (Linn.).

One of our most useful birds is the Starling. They follow the plough in summer with the pied wagtails, feeding on the wireworms ; they are fatal enemies to most of our noxious insects, they breed in holes of trees in our orchards, leaving them with their young just before the cherries get ripe, when they congregate in the marshes, and as the winter approaches vast numbers are seen together probing the turf for every grub they can find, leaving the marks of their bills thick on the grass in all directions. I once had the pleasure of seeing the Starlings leaving the marshes and going to roost ; there must have been hundreds of thousands, strings of Starlings half a mile long following each other in rapid succession for more

than half an-hour ; I wondered where they all came
from.   White varieties are seen every year.

THE ROSE-COLOURED STARLING.

Pastor roseus (Linn.).

It has not been obtained in my district.   One was
shot by a labouring man wide of Maidstone on the
Tonbridge Road in the year 1863.

RAVEN.

Corvus corax, Linn.

The Raven is by no means common, our shepherds
proclaim eternal war against them.

CROW.

Corvus corone, Linn.

Crows are common enough, and frequent trees where
they breed.   They are great pilferers of eggs on the
marsh, and destroy young birds ; nothing seems to
come amiss in the shape of plunder—"black as a
Crow."

GREY CROW.

Corvus cornix, Linn.

The Grey Crow comes in the autumn, feeds along
shore, carries the Mussels and Cockles some height

to drop them on the stone embankments by the river side, and if unbroken rises again with them. Is an indiscriminate feeder, mixes with the Rooks on the newly sown wheat, stays with us till the first week in April, and pecks out the eyes of the first lambs that fall before taking its departure for the north.

ROOK.

Corvus frugilegus, Linn.

See Journ. Board Agriculture Vol
p. 189 re food
67.5 % of their f

Taking my farm as a criterion of the Rooks' usefulness, throughout the summer I see but few, when I begin to sow the wheat in the autumn they come by thousands and continue doing so, off and on, throughout the winter, leaving soon after the barley is sown in the spring. Now if those persons who support a rookery, and where Rooks superabound were to feed them with several sacks of corn per day, they would then know the worth of the depredators they harbour and encourage.

JACKDAW.

Corvus monedula, Linn.

The Jackdaws in winter will sometimes associate with the Rooks, but are not such inveterate thieves, and will make themselves contented with the fresh laid out manure-heaps. I do not, as a rule, find them troublesome. Their breeding haunts are few in my

district ; they used to nest in Rainham Church-tower
before the windows were protected against them by
wire blinds.

MAGPIE.

Pica rustica (Scop.).

Our woods afford a good supply of Magpies.   They
are not very numerous as a species, though constantly
met with.   I am inclined to think they are kept down
by the foxes ; I once saw where a Magpie had been
buried by a fox, leaving the tip end of its tail in
view.   When the weather is rough they roost low,
becoming an easy prey.   The Magpie is a striking
bird in our woodland scenery

JAY.

Garrulus glandarius (Linn.).

As our woods are not preserved, the Jays enjoy
their own freedom, and keeping a good look out are
seldom shot ; a singular thing in their economy is
that we always seem to have exactly the same
number ; it may be that the supply of food is
limited, as our oak trees do not always produce
acorns.   The young broods appear to migrate, or
perhaps like the Magpies are sometimes destroyed
by foxes.   I believe lately our regular number of
Jays may have been interfered with by the cold
wet summers.

GREEN WOODPECKER.

Gecinus viridis (Linn.).

There are no park-like old timber trees in my district to encourage the breeding habits of the Green Woodpecker ; in summer time we never see them, in autumn and winter a few, chiefly immature, get out of bounds and pay us a flying visit ; they are by no means common.

GREAT SPOTTED WOODPECKER.

Picus major (Linn.).

The Great Spotted Woodpeckers, when they do occur, which is very rarely, invariably come in the month of October ; several have been obtained at that time of the year, old birds, both male and female.

LESSER SPOTTED WOODPECKER.

Picus minor (Linn.).

The Lesser Spotted Woodpecker has been met with and obtained on several occasions in our lower orchards adjoining the marsh, a favourite locality both in the autumn and winter. Their appearance, like the Great Spotted Woodpecker's, is very uncertain ; the fine black and white, striped and spotted plumage of each species has been seen, and their tapping of the trees heard with pleasure.

WRYNECK.

Iynx torquilla, Linn.

The Wryneck or " Snakebird " wakes us up early in the spring with its peculiar note, even earlier than the Chiffchaff; the latter frequents the woods and coppices, the former our orchards and homesteads, and enlivens us at once with its familiar presence.

CREEPER.

Certhia familiaris, Linn.

Though pretty generally distributed, the Creeper is not an every-day bird with us ; we now and then catch a glimpse of one in our orchard trees, always climbing upwards from below ; they do upon occasion nest in the old trees.

WREN.

Troglodytes parvulus, Koch.

The note of the Wren seems to be, for the bird's size, one of the loudest of our indigenous birds. Creeping amongst thick herbage, ivy covered walls, stumps of trees, examining every secure spot for insect food in the various stages of life, is fond of roosting in old thatch in cold weather, pulls straws out of the thatch, and frequents the same hole, if undisturbed, year after year.

HOOPOE.

Upupa epops, Linn.

The Hoopoe has upon two occasions at least, been met with in my district, once in the spring of the year, and the other in the autumn, both times near trees on the border of the marsh. A third Hoopoe is said to have been seen this spring, 1894. I have never had the pleasure of meeting with one myself.

NUTHATCH.

Sitta cæsia, Wolf.

As in the case of the Green Woodpecker, we have no forest trees to encourage the Nuthatch. I have only twice seen it, on both occasions in the spring of the year in our orchards ; I was attracted by the very peculiar note or sound which the bird made. Staying about a fortnight it kept repeating the same continuously, and meeting with no response flew away.

CUCKOO.

Cuculus canorus, Linn.

Everybody, when the time of year comes round, is on the look out for the Cuckoo ; they never are or can be deceived ; with us, as a rule, but not invariably, it comes three days after the Nightingale. They are thinly scattered in my district, a pair taking possession of some two miles of country which they appear to keep

to themselves.   Never was bird more welcome ; feeding chiefly upon various species of caterpillers, they
seem to come and go with their favourite insect.   I
once found on brushing round some young trees after
a wet night, a Cuckoo's egg in a half-built Yellow
Bunting's nest ; the egg was stained with the soil of
the field.   I took it away, and in ten days the nest
was finished and furnished with four Yellow Bunting's
eggs.   On another occasion I observed a young
Cuckoo sitting upon almost nothing under the eaves
of a haystack ; it had been reared by a Robin, and
had quite outgrown its compartment.

KINGFISHER.

Alcedo ispida, Linn.

When the Kingfishers appear in my district they
come in the autumn and winter, visit the marsh
dikes, sit upon a barway, stake or any other slight
projection that commands a good view of the water,
waiting for sticklebacks or other small fry upon which
they feed ; on being disturbed flying straight up the
course of the ditch, showing conspicuously the brilliant
colours which render the bird famous.

ROLLER.

Coracias garrulus, Linn.

A Roller was taken alive in our marsh on November 8th, 1888.   My people were employed clamping

mangold wurtzel, when, to their surprise, a Roller
came and settled upon the end of the range where they
were at work, appearing unusually tame; they gave
chase over the hedge, finally catching the bird, which
had taken shelter on the leeward side of a lump of
rushes ; the day was uncommonly cold ; the bird was
in very thin condition, and died soon after it was
caught ; it had received a slight wound, and one
primary was shot asunder.

SWALLOW.

Hirundo rustica, Linn.

The Swallow comes with the Cuckoo and Nightingale in the spring of the year, the one sometimes
appearing before the other. We do not seem to
have so many Swallows now as we used to have, the
successive cold springs and wet summers have told
very much against them ; according to my observation
their numbers have been reduced by more than half.
There was only one pair on my chimney-top last year
instead of three, this may not be the case everywhere ; no doubt with finer summers our insect
loving birds will increase. The spring of the year
1869 was a very severe one for the swallows ; on May
28th, I observed one lying dead beside my farm
buildings and another flying more like a bat than a
bird over my garden hedge ; others were seen by my
brother, Mr. Edward Prentis, at his farm in the parish
of Chalk near Gravesend with their heads under
their wings for warmth, upon his horses' backs in the
stable.

E

MARTIN.

Chelidon urbica (Linn.).

The Martins come in the spring a few days after the Swallows, they are not numerous about here, we have no large colonies, a few scattered nests being all Rainham can boast of. The Martins' white upper tail coverts readily distinguish them from the swallows.

SAND MARTIN.

Cotile riparia (Linn.).

Like the Martin, this species forms no large colonies, as we have no sand banks to speak of. Here and there some spot may contain a nest or two. I once had the pleasure of seeing, though not in this district, about three hundred Sand Martins all in a line upon a field of young turnips, sitting and fluttering along devouring the fly and the turnip beetle.

SWIFT.

Cypselus apus (Linn.)

After the Swallow, the Martin, and the Sand-Martin comes the Swift; were these to start from Africa in a race the last-named would be first. They breed in our Church towers, exercising themselves in a body of a summer's evening, creaking and circling round the steeples till after sunset. They

fly at a greater elevation than the Swallows, it may
be they feed upon another kind of insect, for when
the air, high and low, swarms with flies, they know
their own respective boundaries.

NIGHTJAR.

Caprimulgus Europæus, Linn.

The Nightjar comes to us when our woods are
getting into leaf and the moths are coming out,
which is not before the middle of the month of
May.

On a still summer's evening their rattling notes
can be heard far away. They breed in our three-
years old coppice, choosing the driest spot they can
find, under shelter with an open space above.

I am able to relate a most singular occurrence
which happened to me on the 21st day of June,
1869. When taking a walk at evening time and
towards dusk, adjoining one of our Rainham woods,
I had the pleasure of falling in with a chorus of
notes, quite a novelty to me, lasting for about twenty
minutes, when at nine o'clock, the music gradually
ceased. It was as if all the Nightjars, Cuckoos and
Cricket-birds in the district had assembled together
for a premature and simultaneous departure. The
dew was heavy and the wood struck very cold, so I
wished them all good night and good-bye for
the season; I did not hear or see any of them
again.

Ring Dove or Wood Pigeon.

Columba palumbus, Linn.

The Wood Pigeon with us is not numerous enough to be troublesome; it breeds in our woods sparingly, is most plentiful in winter, feeding upon the turnip-tops, where they are mostly met with and shot.

Stock Dove.

Columba œnas, Linn.

Small flights of Stock Doves are not uncommon, they too, like the Wood Pigeon, visit our turnip fields in winter, are more vigilant, and when keeping together are not easily approached. They sometimes breed in the old cherry trees in my orchards, never in the woods, are fond of an early sown pea field sheltered by the woods and will traverse every inch for a few surface scattered peas.

Turtle Dove.

Turtur communis, Selby.

The Turtle Dove as a rule comes not before the latter end of April, some years it is very plentiful. I have seen thirty or more together feeding on the turnip seed. They frequent the corn-fields where they feed upon the fumitory and other such like seeds. They breed in our woods and are common.

PALLAS'S SAND GROUSE.

Syrrhaptes paradoxus (Pall.).

The year 1863 was the first Sand-Grouse year, six were seen by a shepherd in the Vale of Elmley for several days near a patch of furze; ultimately a shepherd's boy with a gun came upon the scene, and two of them were shot, a male and female; this occurred on the seventh of June, they happened to come into my possession. Four Sand-Grouse were seen flying in a westerly direction two days afterwards, when another, a female, was shot by a gamekeeper.

The year 1888 was the second Sand-Grouse year, four were seen upon a ploughed field for several days in the parish of Hoo; on December the fourteenth a male was picked up dead, with head cut clean off by the telegraph wires on the Isle of Grain railway, another I believe was shot about the same time near Sheerness.

PHEASANT.

Phasianus colchicus (Linn.).

Pheasants are more or less plentiful according as they are preserved.

PARTRIDGE.

Perdix cinerea, Lath.

The well-known, highly esteemed Partridge is generally distributed in the district. Our chalk soil

with hill and valley, field and wood combined, makes
excellent sport for the first of September.

A white or cream-coloured variety has been shot.

RED-LEGGED PARTRIDGE.

Caccabis rufa (Linn.).

The red-legged Partridge has spread within these
few years all over my district ; they are fine showy-
looking birds and in every way capable of taking
care of themselves. Their habit of running before
the dogs causes them to be of little use to the sportsman.

QUAIL.

Coturnix communis, Bonnaterre.

The Quail, when inclined to pay us a visit, always
does so in the spring of the year, about the middle of
May. Their loud clear notes are occasionally heard
in our corn and clover fields, where they sometimes
breed, they are not an everyday bird ; several years
may pass without one being heard or seen.

GREAT OR NORFOLK PLOVER.

Œdicnemus scolopax (Gmel.).

The Great Plover must be considered scarce in the
district ; one was shot December 23rd, 1886, another

in the month of March, 1892; one only has occurred to me upon a fallow field in the month of June. A few others have been met with.

GOLDEN PLOVER.

Charadrius pluvialis, Linn.

November is the month for the Golden Plover, their arrival is most uncertain, some years they are tolerably plentiful, others very scarce. They frequent the plough fields, fly together in company of an evening over the low hedges, where they may be intercepted and shot. Should they not migrate until May, which sometimes does occur with a few, their black breasts are then conspicuous.

DOTTEREL.

Charadrius morinellus (Linn.).

Dotterels have occurred on the Isle of Elmley, May 3rd, 1870, and again on May 19th, 1876. A young bird was obtained on the Isle of Sheppey, Sep. 18th 1864, all of them out of the bounds of my district.

RINGED PLOVER.

Ægialitis hiaticula (Linn.).

The Ringed Plover is the first river shore bird which I have to note. Owing to the great amount of traffic on the Medway they are somewhat thinly scattered but, notwithstanding the noise of vessels, our river flats are never without Ring Plover in the autumn and winter.

GREY PLOVER.

Squatarola helvetica (Linn.).

The Grey Plovers make their appearance on the Medway in the month of November, our sportsmen are then fully on the alert, and those which the vessels do not drive away generally fall into their pockets.

PEEWIT.

Vanellus vulgaris, Bechs.

Our marshes are never without Peewits in the spring of the year, where they lay their eggs and sometimes breed ; they flock together on our ploughed fields in the autumn, doing much good by feeding upon the snails and probably other insect pests, but I am sorry to say they are much persecuted.

TURNSTONE.

Strepsilas interpres (Linn.).

A few Turnstone Plovers come on the Medway in the autumn and are met with sparingly in winter ; on one occasion a specimen was procured August 1892, in summer plumage.

OYSTER CATCHER.

Hæmatopus ostralegus, Linn.

The Oyster Catcher seldom comes on the Medway, there is a great deal too much traffic going on for them ; a few appear singly and are sometimes shot.

HERON.

Ardea cineria, Linn.

Herons visit our marshes and the river Medway all the year round, most numerous in the autumn when the young are able to fly ; they come from a neighbouring heronry at Cobham, near Gravesend, where they are strictly preserved by the Earl of Darnley.

PURPLE HERON.

Ardea purpurea, Linn.

An immature specimen of the Purple Heron was shot a few years ago in the vicinity of an old decoy-pond where some patches of reeds grow, near the Swale river.

BITTERN.

Botaurus stellaris (Linn.).

Our forty acres of spring marshes, where long thick herbage grows in the spring-water ditches adjoining some plantations, is a spot peculiarly adapted for the Bittern, and where they have been met with and shot upon at least four occasions ; others also have occurred in the district.

SPOONBILL.

Platalea leucorodia, Linn.

A fine adult Spoonbill, with a buff collar and pendant crest, was shot on the Isle of Elmley, April 12th 1865. Immature specimens have been met with on three occasions on the marshes near the river Medway.

BLACK STORK.

Ciconia nigra (Linn.).

In July 1884 a Black Stork was seen to visit an island marsh ; the water in the ditches after some very dry weather became nearly dried up, leaving only a few puddles in places where the eels had collected ; a trap was set by the shepherd, who instead of catching an eel caught the Black Stork, and nothing would do but that he must eat it. On July 24th, myself and Mr. Chas. Gordon, of the Dover Museum happened to be walking along a creek and when opposite the Shepherd's house we picked up the remains of a Black Stork, after it had floated backwards and forwards with the tide for some days, it was in a very mutilated condition, but we managed to secure the scull, feet and pinions as proof positive.

Another Black Stork was afterwards seen, and in all probability it shared a similar fate.

CURLEW.

Numenius arquata (Linn.).

Curlews frequent our salt-marshes and mud-flats in considerable numbers, they make their appearance the latter end of August; our shooters lose no time in taking advantage of their arrival; with dogs for the purpose, they conceal themselves along shore, the Curlews, supposing the dog to be a fox, immediately give chase, uttering their peculiar cry and are shot. Some few Curlews remain with us throughout the summer, but they do not attempt to breed.

WHIMBREL.

Numenius Phœopus (Linn.).

The seventh day of May is the grand time for the Whimbrels. They are distributed singly on the edge of the river Medway, and by the side of every creek, following the tide as it recedes, returning with the tide when it flows, and after staying a week or ten days they are all off together, not a single one being left behind. In the autumn they make no stay, flying high overhead we hear their clear whistle.

REDSHANK.

Totanus calidris (Linn.).

The Redshank is an old inhabitant of our salt-marsh and mud-flats, probably from time immemorial; there they constantly breed and rear their young. Owing, however, to the extended traffic on the Medway and the carrying away of the marsh clay for the purposes of cement, their numbers in the breeding-season are becoming gradually reduced; some few still favour us in the spring with their loud ringing call and roundabout flight, reminding us, wherever a piece of salt-marsh remains, of their close attachment to the spot.

SPOTTED REDSHANK.

Totanus fuscus (Linn.).

The Spotted Redshank pays our mud-flats a visit generally in the autumn. On one afternoon I met with a pair in the front of a snow-storm in mid-winter. Young birds appear to be the rule, old birds the exception. The latter are met with occasionally.

GREEN SANDPIPER.

Totanus ochropus (Linn.).

Common enough in the autumn about September and October. In that wet year, 1860, a pair remained throughout the summer; they do not feed on our mud flats like most of the other Sandpipers, but appear to confine themselves to the marsh ditches, probably attracted by the small fry and insects that abound in such places. On being disturbed their rapid flight, white rumps and loud whistle enlivens the marsh.

WOOD SANDPIPER.

Totanus glareola (Gmel.).

It was after a storm of thunder and lightning and a tremendous heavy rain at night, which happened on the 26th July, 1867, that on the following day a flock of about one hundred Wood Sandpipers appeared in our marsh, five of them were shot, other single birds were afterwards met with.

COMMON SANDPIPER.

Totanus hypoleucus (Linn.).

More common on their return journey in the autumn than their forward northern journey in the spring, when their visits are few. In the former season they frequent our marsh ditches with the Green Sandpiper, but do not appear to associate with it. On being disturbed the latter flies high, the Common Sandpiper flies low along the water and round the bend of the ditches; their stay is not so prolonged as that of the Green Sandpiper.

AVOCET.

Recurvirostra avocetta, Linn.

The Avocet is extremely rare; only upon one occasion, to my knowledge, having been met with, which was on the 23rd September, 1887, one was seen flying along a creek and shot; a bird of the second year; another was observed soon afterwards, which I believe escaped.

BLACK-TAILED GODWIT.

Limosa ægocephala (Linn.).

In the month of January, 1881, several immature Black-tailed Godwits were met with, flying along the edge of our Medway saltings at the flow of the tide, one of them was shot. October 20th, 1882. A pair of adult Black-tailed Godwits, accompanied with a pair of adult Spotted Redshanks, were together on the border of one of our Medway creeks; one of each was shot.

RUFF.

Machetes pugnax (Linn.).

Adult Ruffs on their migratory passage are extremely rare; immature birds in the autumn are frequently obtained, they seem to prefer the marsh ditches to the more open banks of the Medway.

BAR-TAILED GODWIT.

Limosa lapponica (Linn.).

A common autumn migrant, more numerous on the mud flats and the banks of the Medway some seasons than others. I have never known them to be seen on the spring passage in their red plumage, probably they take a direct course along the coast rather than coming inland.

WOODCOCK.

Scolopax rusticula, Linn.

This inestimable bird,—a general favourite with every sportsman,—is not very numerous as a species with us, our woods lying upon the chalk seem to be too dry for their requirements ; still we are never without a few, and the most I have heard of being shot in a day is a brace. Some scattered single birds are all we can boast of. In the spring of the year it may be a few others are seen on passage, flying over the woods on the eve of their departure for the far north.

GREAT SNIPE.

Gallinago major (Gmel.).

Has not to my knowledge been met with, but I cannot help thinking that it must have occurred.

COMMON SNIPE.

Gallinago cælestis (Frenzel).

The shooting of the Common Snipe, as proof of a sportman's skill is not practised here so often as could be wished. About half a dozen shots in a day is the most our dry marshes will afford ; they come and go, as every sportsman knows, here to-day and gone to-morrow, and often does it occur that when most expected they are never seen at all.

JACK SNIPE.

Gallinago gallinula (Linn.).

Our Rainham marsh usually contains in the winter season two or three Jack Snipes ; they come, as a rule, the latter end of November. I once shot a " Jack " at the beginning of April in the plumage of glossy green.

CURLEW SANDPIPER.

Tringa subarquata (Güldenstädt).

The Curlew Sandpipers come the beginning of September. They frequent the shores of our creeks in small numbers as a species, and are very tame. Once, while rowing a boat, a pair came and settled some five yards off me. They sometimes occur in the half changed plumage from summer to winter ; their stay is about a month.

KNOT.

Tringa canutus, Linn.

Immature Knots visit our Medway mud flats in very small numbers in the autumn ; half a dozen is the most I have met with at the same time together ; others are driven into our creeks in severe weather.

LITTLE STINT.

Tringa minuta, Leisler.

The Little Stint is extremely rare as a species on the river Medway. A pair was obtained the 17th day of September, 1881 ; they were shot flying along a ditch on an island marsh, and another pair or two have been shot on the margins of our creeks.

TEMMINCK'S STINT.

Tringa Temmincki, Leisl.

Temminck's Stint has occurred on more than one occasion on our Rainham marsh. I once had the pleasure of shooting the bird, but this was outside my district, on the Isle of Sheppey ; it was flying the opposite side of a wide ditch ; I shouldered my gun thinking to shoot a Dunlin, and on pulling the trigger I observed the bird was smaller. I picked up a Temminck's Stint in winter plumage the 1st November, 1869.

F

DUNLIN.

Tringa variabilis, Linn.

The Dunlin is more or less numerous on the banks of the Medway and mud-flats according to the severity of the season ; I have seen them in fifties, in hundreds, and in thousands ; they begin to make their appearance in August, when other shore birds, such as ringed Plovers, consort with them ; I have shot them flying together with jet-black breasts ; we have a smaller variety which I have met with on the borders of our creeks in the month of April after all the rest of the Dunlins are gone, in red plumage and with less black on the breast, they appear to be a distinct race. I have not heard of their making any prolonged stay with us.

PURPLE SANDPIPER.

Tringa striata, Linn.

When the winter has been severe I have shot the Purple Sandpiper on our marsh walls, where they have been repaired with rock-stone.

RED-NECKED PHALAROPE.

Phalaropus hyperboreus (Linn.).

A red-necked Phalarope has been obtained in the autumn time of the year on the Rainham marsh, another was shot swimming on the Isle of Elmley, November, 1867, both immature.

GREY PHALAROPE.

Phalaropus lobatus (Linn.).

This neat, chaste, pretty bird does now and then pay us a visit, but their visits are few, only two or three to my knowledge have been obtained on the banks of the Medway in my district, others possibly may have occurred.

LANDRAIL.

Crex pratensis, Bechs.

September is our month for the Landrail; they resort to the clover-fields where the sportsman is pretty sure to find them. On one occasion, when all the young clover seeds failed to grow, I happened to leave my clover field for another year, being the only one in the neighbourhood, I shot twenty-five Landrails on nine acres in September, this was, however, a most unusual event; they commonly are met with singly, or at most in twos or threes. On one occasion I shot a Landrail in the month of December in a turnip-field.

SPOTTED RAIL.

Poranza maruetta (Leach).

Our Rainham marsh is not always without a Spotted-Rail; when they appear it is in the autumn, generally in the month of October ; several have been obtained and others from their creeping habits have, no doubt, passed unnoticed.

WATER RAIL.

Rallus aquaticus (Linn.).

Almost every winter our marsh ditches and spring water-courses contain a Water Rail or two, the sportsman in quest of Snipe meets with them, their laboured flight affords an easy shot.

MOOR HEN.

Gallinula chloropus (Linn.).

Having only a small extent of spring or fresh-water marsh at Rainham, the Moor Hen is not with us a very common resident ; where thick herbage grows a pair or two take up their abode, and in the absence of persecution rear their young, they have been picked up dead on our railway, having flown against the telegraph wires on migratory passage.

COOT.

Fulica atra, Linn.

Seldom or never seen on our Rainham marshes. I know of only one instance of their being observed ;

a pair came, built a nest and laid an egg, which was immediately destroyed by a carrion crow, in a bunch of rushes near the outside of a broad piece of water, they took their departure and were no more seen. They sometimes occur on passage near woods and other out-of-the-way places.

WILD GREY GEESE AND WILD SWANS.

*

Owing to the excessive traffic on the Medway from the constant sailing to and fro of vessels to the number of two hundred every day, the wild Grey Geese and Wild Swans have no resting place there. My acquaintance is limited to seeing them fly over at a considerable height in very severe weather, first to the south, afterwards back again to the north ; merely their outline is visible.

BRENT GOOSE.

Bernicla brenta (Pall.).

The Brent Geese come into our creeks in severe winters in very small numbers ; no sooner do they appear than they speedily have notice to quit and are off, one or two are sometimes shot.

* I have had no means of identifying the species ; they are either Geese or Swans.

BEWICK'S SWAN.

Cygnus Bewicki, Yarrell.

On January 22nd, 1879, an excellent specimen of
Bewick's Swan was shot in an extraordinary way.
Some shooters were lying in wait for sea-gulls under
the shelter of a wood, the gulls returning from the
fields which had been manured with sprats, when a
Swan came flying in the same direction, following the
track of the Gulls at some height ; a number six shot
happened to touch a wing, the bird immediately
lowered, and dropped about a quarter of a mile off ;
it was pursued and captured.

SHELDRAKE.

Tadorna cornuta (Gmel.).

The Sheldrake comes into our river in severe
winters, always leaving the broad water to the many
vessels of trade ; they drop like other fowl under the
leeward side of a few sheltered spots in our creeks.

SHOVELLER.

Spatula clypeata (Linn.).

The noise and bustle on our river keeps nearly all
the wild fowl away, the Shoveller does not appear to
be one of those which will easily submit to it, they
come and are off again ; are very seldom shot.

Gadwall.

Anas strepera, Linn.

The Gadwall taking a view of our numerous craft gives them a wide berth ; are very seldom if ever shot.

Pintail.

Dafila acuta (Linn.).

The Pintail is much more common than the two preceding species. Scarcely a severe winter passes without a pair or two being shot on some of the creeks and salt marshes on the Medway.

Gargany.

Querquedula circia (Linn.).

A pair of Garganys were shot on one of our island marshes on the 7th day of March, 1874, and three immature birds were shot on one of our creeks, August, 1893.

Wild Duck.

Anas Boschas, Linn.

In severe winters after a strong breeze from the east, Wild Ducks in small numbers come into our creeks for shelter. A neighbouring proprietor has lately enclosed a wide extent of salt marsh, converting it by means of embankments into a pasture marsh for sheep, and from philanthropic motives has succeeded in establishing a pond in the centre for the preservation of Wild Duck ; they separate and pair in March, are seen to frequent most of the adjacent marshes where they breed.

A note from me was published in the *Zoologist*, and copied into many local and other newspapers, as follows :—

" A mowing machine was set to work, June, 1891, round the outside of a field of lucerne bordering our marsh, diminishing the circle each time round the field, leaving about two acres in the centre. A Wild Duck was seen by the shepherd to fly from the piece of lucerne that was left with something in her beak, and happening to fly near him, she dropped a three parts incubated egg. She was observed by the shepherd and also by the sheep-shearer carrying another egg in her beak, this time over the marsh wall towards the saltings, and again she was seen, the third time, carrying an egg in her beak in the same direction. On the mowing machine going to work the next day, and finishing the field by mowing the last piece of lucerne, the Wild Duck's nest was discovered from which the eggs had been removed."

A Wild Duck has been met with transporting her newly-hatched brood upon her back across the Medway.

TEAL.

Querquedula crecca (Linn.).

This pretty little Duck is not so numerous as we could wish, being very thinly distributed on our marshes ; they are rather more plentiful of late, particularly in the vicinity of the duck-pond.

WIGEON.

Mareca Penelope (Linn.).

On the Medway are to be seen generally more vessels than birds ; how then is it possible for Wigeons to gain a settlement for the shortest period ? They come, are off again, perhaps even faster than they came ; it must be a very severe winter for a few to drop on our creeks.

EIDER DUCK.

Somateria mollissima (Linn.).

A pair of Eider Ducks were shot on the Medway a few years ago in their immature brown plumage.

VELVET SCOTER.

Œdemia fusca (Linn.).

Velvet Scoters do sometimes appear on the Medway, for the most part either females or immature ; several have been shot.

COMMON SCOTER.

Œdemia nigra (Linn.).

The Common Scoter has not, to my knowledge, been shot on the Medway.

POCHARD.

Fuligula ferina (Linn.).

We have no harbour for the Pochard, come it does upon occasion and goes, like Wigeon, even faster than it comes. The incessant traffic on our river, which increases every year, bids fair to drive the Pochard right away.

SCAUP DUCK.

Fuligula marila (Linn.).

The Scaup Duck is certainly much less uncommon than the Pochard ; at uncertain intervals, mostly in severe weather, our gunners obtain a few, as a rule chiefly single birds, numbers being quite the exception.

TUFTED DUCK.

Fuligula cristata (Leach).

The Tufted Duck is not uncommon. They come into the creeks and marshes for the most part singly, and always in winter time ; are thinly distributed.

LONG-TAILED DUCK.

Fuligula glacialis (Linn.).

Two examples of the Long-tailed Duck have come to my knowledge as shot on the Medway, both specimens immature, others probably may have been obtained.

GOLDEN EYE.

Clangula glaucion (Linn.).

I know of but one instance of the mature Golden Eye being shot on our river; young birds are frequently shot, and on one occasion I saw a small flock of Golden Eyes, seven in number, one being almost white; on the first sign of danger they were off.

SMEW.

Mergus albellus, Linn.

Tuesday, the 18th day of January, 1881, will ever be remembered for the most tremendous storm of wind and snow from the east which has been experienced in this country during, perhaps, the present century. This had the natural effect of driving every thing before it. The Wild Fowl on this occasion were no exception, blowing them into our river and creeks for shelter in every direction. On the weather calming down, a pair of Smews were seen; one of them, a female, was shot. After a few hours rest, all

the wild fowl departed, and the vessels again began
to sail, our river resuming its normal condition.

A beautiful male Smew was shot on one of our
island marshes a few years previously.

RED-BREASTED MERGANSER.

Mergus serrator, Linn.

The Red-breasted Merganser is not uncommon, it
comes into our creeks on the Medway whenever the
weather is cold enough, it appears to be partial to
the creeks, never flying by choice over the land, has
been in many instances shot, fine handsome adult
males, also females, but the young are the most
numerous.

GOOSANDER.

Mergus merganser, Linn.

I only know of one instance of the Goosander
coming on the Medway ; an immature bird was shot
while flying at some height over our saltines in
winter time a few years ago.

GREAT CRESTED GREBE.

Podiceps cristatus (Linn.).

The Great Crested Grebe make its appearance
singly nearly every winter on our creeks. They are

somewhat difficult to shoot owing to their diving powers, but are now and then obtained.

RED-NECKED GREBE.

Podiceps griseigena (Bodd.).

The red-necked Grebe comes into our creeks at uncertain intervals in winter time or early spring, a few specimens have been obtained.

SCLAVONIAN GREBE.

Podiceps auritus (Linn.).

The Sclavonian Grebe comes in winter, is not very common, but perhaps more so than all the rest of the Grebes, excepting the little Grebe or Dabchick ; and it appears to frequent the open river in preference to the creeks.

EARED GREBE.

Podiceps nigricollis, Brehm.

I have only known of a single pair of Eared Grebes being shot, both immature, in the month of September, 1881. They are certainly the most rare of all the Grebes on the Medway.

Little Grebe.

Podiceps fluviatilis (Tunstall).

The Little Grebe is more numerous than all the rest of the Grebes put together; it goes by the name of the Dabchick, frequents our marsh ditches in summer, on the borders of which it breeds, always on the water adjoining a bunch of rushes or solitary bush or other such like shelter, builds a compact nest of wet herbage; the eggs are never left exposed, being concealed from view by the parent on leaving her nest. In winter the Dabchick with us is more scarce and but seldom seen.

Great Northern Diver.

Colymbus glacialis (Linn.).

Young, immature Great Northern Divers are sometimes met with and shot on the Medway.

Black-throated Diver.

Colymbus arcticus, Linn.

Immature Black-throated Divers are also, like the previous species, occasionally met with and shot on our river.

Red-throated Diver.

Colymbus septentrionalis, Linn.

The red-throated Diver, locally called Sprat Loon, is common on the Medway, observed in the winter and spring with speckled back and white throat, most numerous in the month of March. I once happened to shoot a Sprat Loon from a boat, the day was clear, the tide low, the water smooth as glass. After falling it dived out of sight, then protruding its beak only above water for air floated and dived away.

Guillemot.

Uria troile (Linn.).

The Guillemot is not very common on the Medway, come it does at intervals and is sometimes shot.

Black Guillemot.

Uria grylle (Linn.).

I am not aware that the Black Guillemot has ever appeared on the Medway.

Little Auk.

Mergulus alle (Linn.).

A storm-driven specimen of the Little Auk may have occurred on or near the Medway.

PUFFIN.

Fratercula arctica (Linn.).

The Puffin is extremely rare on the Medway, I only know of two being shot. A storm driven Puffin was picked up dead on our marsh after the November gale of 1893.

RAZOR-BILL.

Alca torda, Linn.

The Razor-bill has not to my knowledge occurred on the Medway; it may possibly have done so following the track of the Guillemot.

CORMORANT.

Phalacrocorax carbo (Linn.).

It is only since the last five or six years that the Cormorant has become plentiful on our river; it may be that they are getting more numerous round the coast. They do not appear to be intimidated by the sight of the many vessels sailing to and fro all day and night, as they constantly fly past them and some-times get shot.

SHAG.

Phalacrocorax graculus (Linn.).

The Shag, or Lesser Cormorant has to my knowledge been observed on the Medway and shot on one
or two occasions.

GANNET.

Sula Bassana (Linn.).

Our river affords no resting place for the Gannet,
which scarcely ever comes beyond the line of vessels.
A few years ago a Gannet was found near the centre
of our wood. A flock of Rooks were seen by a
labourer circling round and cawing vociferously over
the spot where a dead Gannet was discovered which
had evidently been storm-driven.

CASPIAN TERN.

Sterna caspia, Pall.

Some few years ago, before the river became so
thickly studded with vessels, I had the pleasure one
autumn of meeting with the Caspian Tern; it frequented a wide part of the Medway, and an adjoining wide creek for about a fortnight; happening to
shoot a Common Tern from my yacht, the great bird
came, hovered for a second or two over it, but from
the distance the little bird had floated, I was unable
to procure the Caspian Tern.

G

Sandwich Tern.

Sterna cantiaca, Gmel.

The Sandwich Tern is but seldom seen on the Medway. It comes in the autumn when a few immature specimens have been shot.

Common Tern.

Sterna fluviatilis, Naum.

This is the Common Tern or Sea Swallow of the Medway ; provided the season is favourable they sport on the calm water in the autumn for a fortnight at least, their graceful pitching and rapid flight is always pleasing to see.

Arctic Tern.

Sterna macrura, Naum.

I have never been able to recognise the Arctic Tern among the many common Terns that sometimes visit our river ; in the autumn possibly they may have occurred, but have not to my knowledge been shot.

Lesser Tern.

Sterna minuta, Linn.

The Lesser Tern is not uncommon on the Medway.

In the month of August, at which time of year I have generally met with them, they visit our creeks in small numbers.

BLACK TERN.

Hydrochelidon nigra (Linn.).

Like the Lesser Tern, the Black Tern visits us in the month of August; I have met with them sitting upon the buoys on the Medway, and never in any other place, always immature.

LITTLE GULL.

Larus minutus, Pall.

February 7th, 1870, I had the pleasure of meeting with and shooting the Little Gull flying over a plough field bordering a marsh, when it was blowing a strong wind from the east, it was in the second year's plumage; in March I saw another Little Gull with a black head accompanying a flock of the Common Black-headed Gulls up a creek, it separated and flew towards and over a flock of sheep that were grazing on a marsh bank.

February 14th, 1874, an adult Little Gull was shot at the mouth of Milton Creek.

September 17th, 1884, a young immature Little Gull was shot in mottled plumage on our river.

G 2

BLACK-HEADED GULL.

Larus ridibundus, Linn.

The Black-headed Gull is common enough on the
Medway in the autumn, winter, and early spring, and
when they begin to assume their black heads they
depart ; visit our sprat fields, coming by hundreds ;
should the weather be cold they follow the waggons
from the wharf into the field ; only by continually
shooting are they kept away, they wait in an adjoining field at a little distance in the line of scent,
watching an opportunity in order to pick up a few
sprats in an unprotected corner ; should the gunner
be absent for a few minutes down they come ready to
carry all the sprats away ; when evening arrives they
retire to the river where they roost.   The following
day they repeat their excursions to the sprat field,
stealing some if they can till the sprats are all
ploughed in.   Should a frost intervene to stop the
plough the Gulls are again on the alert, not losing
a chance all the time a sprat is to be had.   The
Black-headed Gull in rough weather often visits the
fresh turned ploughed furrows, following the plough
across the field, frequently in company with rooks,
robbing the land of the earth-worms.

KITTIWAKE.

Rissa tridactyla (Linn.).

The Kittiwake is rather a rare straggler on the
Medway and seldom met with.   The few that are

shot are generally immature birds ; incidentally an adult may be shot on the river or flying over the land.

COMMON GULL.

Larus canus, Linn.

The Common Gull is of frequent occurrence on the Medway ; they are pretty generally distributed, more particularly where food is to be had, it never comes inland.

LESSER BLACK-BACKED GULL.

Larus fuscus, Linn.

The Lesser Black-backed Gull comes on the Medway in small numbers, flies high, therefore the black back and white breast is plainly distinguished at a distance ; the immature, with brown mottled plumage, are often shot.

HERRING GULL.

Larus argentatus, Gmel.

The Herring Gull flies over the land at the time of the sprat season and makes familiar acquaintance with the Black-headed Gull ; like them it follows the waggons from the wharf into the field, keeping them company all the while the sprats are exposed, they are not so numerous as the black-heads by about 30 to 1 ; in every respect they continue with each other until the coast is cleared and the sprats all covered in by the plough.

GREAT BLACK-BACKED GULL.

Larus marinus, Linn.

The great amount of traffic on the Medway keeps the Great Black-backed Gull at a distance ; they do not like the noise and bustle, therefore make themselves scarce ; are sometimes shot, most often in immature plumage.

GREAT SKUA.

Stercorarius catarrhactes (Linn.).

The Great Skua has not to my knowledge ever appeared on the Medway.

POMERINE SKUA.

Stercorarius pomatorhinus (Temm.).

I must extend the bounds of my district in order to include a pair of Pomerine Skuas, which were shot by a bargeman when in the act of killing a Common Gull, beyond Rochester bridge, the 20th February, 1882, one a fine adult with yellow collar and a long turned up tail, the other immature, in the brown plumage. A Pomerine Skua was shot flying

over Chatham Hill in windy weather, November
27th, 1890, either in the second or third year's
plumage.

ARCTIC SKUA.

Stercorarius crepidatus (Gmel.).

The Arctic Skua is of frequent occurrence on the
Medway in the months of September and October ;
they are seen chasing the Common Tern and the
Black-headed Gull ; they are always in the immature
plumage, the colour which prevails is a dark chocolate
brown. Only on one occasion has another variety
been met with, which is spotted sandy and brown.
The mature bird with a long tail is extremely scarce,
one was shot out of the bounds of my district,
October, 1865, on the Swale River.

LONG-TAILED SKUA.

Stercorarius parasiticus (Linn.).

I have heard of no specimen of the Long-tailed
Skua ever having been shot on the Medway.

FULMAR PETREL.

Fulmarus glacialis (Linn.).

Exactly the same may be said of the Fulmar
Petrel, which I believe has never been seen on the
Medway.

Forked-Tailed Petrel.

Cymochorea leucorrhoa (Vieill.).

The Forked-tailed Petrel has occurred at least on one occasion in the neighbourhood of the Medway.

Stormy Petrel.

Procellaria pelagica, Linn.

A Shepherd's boy hearing a squeaking in the air at the same time descending from some height, picked up a Stormy Petrel alive.

THE END.

INDEX.

Woodfall & Kinder, Printers, 70 to 76, Long Acre, London. W.C.

Books of Local Observation

ON BIRDS.

The Birds of Lancashire. By F. S. MITCHELL. Second Edition. Revised and Annotated by HOWARD SAUNDERS, F.L.S. &c., with additions by R. J. HOWARD, and other local authorities. 297 pages, demy 8vo, with map and 12 illustrations, 10s. 6d.

The Birds of East Kent, A Tabulated List and Description of, with Anecdotes and an Account of their Haunts. By GEORGE DOWKER, F.G.S. 8vo, sewed, 2s. 6d.

The Birds of Middlesex. By J. E. HARTING. Post 8vo, 7s. 6d.

The Birds of Somersetshire. By CECIL SMITH. Post 8vo, 7s. 6d.

The Birds of Norfolk. By the late HENRY STEVENSON. Completed by THOMAS SOUTHWELL. 3 vols. 8vo, £1 11s. 6d.

The Birds of the Humber District. By JOHN CORDEAUX. Post 8vo, 6s.

The Birds of Suffolk. By CHURCHILL BABINGTON, D.D., V.P.R.S.L., &c. 8vo, cloth, 10s. 6d.

Bird-Life of the Borders: Records of Wild Sport and Natural History on Moorland and Sea. By ABEL CHAPMAN. With numerous Illustrations by the Author. 8vo, 12s. 6d.

The Birds of Jamaica. By P. H. GOSSE. Post 8vo, 10s.

The Birds of Egypt. By Captain G. E. SHELLEY, F.Z.S., &c. Royal 8vo., Coloured Plates, £1 11s. 6d.

The Birds of Damara-Land and Adjacent Countries of South-West Africa. By the late C. J. ANDERSSON. Arranged and edited, with Notes by JOHN HENRY GURNEY. 8vo, 10s. 6d.

The Birds of Rainham, including the District between Chatham and Sittingbourne. By WALTER PRENTIS. Post 8vo.

GURNEY & JACKSON, 1, Paternoster Row.
(MR. VAN VOORST'S SUCCESSORS.)

OTHER BOOKS ABOUT BIRDS

PUBLISHED BY

GURNEY AND JACKSON.

A History of British Birds. By the late WM. YARRELL, V.P.L.S.,
F.Z.S. Fourth Edition, revised to the end of the Second Volume
by Professor NEWTON, M.A., F.R.S. The revision continued by
HOWARD SAUNDERS, F.L.S. 4 vols. 8vo, cloth, with 564 Illustrations, £4.
> "The Fourth Edition of 'Yarrell' will remain for many years a classic
> without a rival."—*Academy.*

The Fowler in Ireland: or, Notes on the Haunts and Habits of
Wild Fowl and Sea Fowl, including Instructions in the Art of
Shooting and Capturing them. By Sir RALPH PAYNE-GALLWEY,
Bart. With many Illustrations of Fowling, Birds, Boats, Guns,
and Implements. 8vo, £1 1s.

The Book of Duck Decoys, their Construction, Management, and
History. By Sir RALPH PAYNE-GALLWEY, Bart. Crown 4to,
cloth, with coloured Plates, Plans, and Woodcuts, £1 5s.

Notes on Sport and Ornithology. By His Imperial and Royal
Highness the late CROWN PRINCE RUDOLF OF AUSTRIA. Trans-
lated, with the Author's permission, by C. G. DANFORD. Demy
8vo, 650 pages, with an Etching by FRANK SHORT, 18s.

Wild Spain. By ABEL CHAPMAN and WALTER J. BUCK. 8vo, with
174 Illustrations, £1 1s.
> "The book is a prize to all who are interested in Spain, or in the pursuit
> and study of wild game. To read it is to be smitten with the desire to seek
> out the flamingo in the marshes, the bustard and the wild boar in the cornlands
> and cork-oak thickets, the ibex and the lammergeier in the high Sierras."—
> *National Observer.*

A Handbook of British Birds. Showing the Distribution of the
Resident and Migratory Birds in the British Islands, with an
Index to the Records of the Rarer Specimens. By J. E. HART-
ING, F.L.S, &c. 8vo, 7s. 6d.

Hints on Shore-Shooting, including a Chapter on Skinning and
Preserving Birds. By J. E. HARTING, F.L.S. Post 8vo, 3s. 6d.

A List of British Birds. Compiled by a Committee of the British
Ornithologists' Union. 8vo, sewed, 10s. 6d.

A List of British Birds. Second Thousand. Revised by HOWARD
SAUNDERS, F.L.S., F.Z.S., &c. For labelling Specimens or for
Reference. 8vo, sewed, 6d.

GURNEY & JACKSON, 1, Paternoster Row.

(MR. VAN VOORST'S SUCCESSORS.)

In One Volume, 750 pages, demy 8vo, cloth, with 367
fine Woodcuts and 3 Maps, £1 1s.

AN

ILLUSTRATED MANUAL

OF

BRITISH BIRDS.

BY HOWARD SAUNDERS, F.L.S., F.Z.S., &c.

Editor of the Third and Fourth Volumes of the Fourth Edition of
" Yarrell's History of British Birds."

" It would be difficult to give a better condensation of
facts in fewer lines than has been contrived by Mr.
Saunders."—*Zoologist.*

" Perhaps the question most frequently put to a zoologist
by a lay friend is, ' What is a really good book on British
Birds that is not too expensive?' and the question has been
one that has been found extremely difficult to answer.
Mr. Saunders deserves our thanks for having taken this
difficulty out of our way."—*Athenæum.*

" Excellent alike in style and matter, it ought to be in
the hands of every lover of birds, and should take the place
of several inferior books on the subject now before the
public."—*Annals of Natural History.*

" It is scarcely necessary to inform those who are
acquainted with the previous work of the author, that the
information is not only valuable from its correctness, but
that it is brought up to the present date."—*Field.*

GURNEY & JACKSON, 1, Paternoster Row.
(MR. VAN VOORST'S SUCCESSORS.)

STANDARD WORKS

ON

BRITISH NATURAL HISTORY.

The Natural History and Antiquities of Selborne. By the late Rev. GILBERT WHITE. Edited by THOMAS BELL, F.R.S., &c. 2 vols. 8vo, with Steel-plate and other Illustrations, £1 11s. 6d. (A few copies large paper, royal 8vo, with the Plates on India Paper, £3 3s.)

A History of British Quadrupeds; including the Cetacea. By Professor BELL, F.R.S., &c. Second Edition, revised and partly rewritten by the Author, assisted by R. F. TOMES and E. R. ALSTON. 8vo, illustrated by 160 Woodcuts, £1 6s.

History of British Reptiles. By Professor BELL. Second Edition, with 50 Illustrations, 12s.

Yarrell's History of British Fishes. Third Edition, with Figures and Description of the additional Species by Sir JOHN RICHARDSON, C.B., and with a Portrait and Memoir. 2 vols. 8vo, 522 Illustrations, £3 3s.

History of British Stalk-Eyed Crustacea (Lobsters, Crabs, Prawns, Shrimps, &c.). By Professor BELL. Illustrated by 174 Engravings. 8vo, £1 5s.

History of British Sessile-Eyed Crustacea (Sandhoppers, &c.). By C. SPENCE BATE, F.L.S., and J. O. WESTWOOD, F.L.S., &c. Many Illustrations. 2 vols. 8vo, £3.

History of British Mollusca and their Shells. By Professor EDWARD FORBES, F.R.S., &c., and SYLVANUS HANLEY, B.A., F.L.S. Illustrated by 203 copper-plates, 4 vols. 8vo, £6 10s.; royal 8vo, with the Plates Coloured, £13.

A History of the British Hydroid Zoophytes. By the Rev. THOMAS HINCKS, B.A. 2 vols. 8vo, cloth, with 67 Plates, £2 2s.

History of the British Zoophytes. By GEORGE JOHNSTON, M.D., LL.D. Second Edition, in 2 vols. 8vo, £2 2s.

History of British Starfishes and other Animals of the Class Echinodermata. By Professor EDWARD FORBES. 8vo, 120 Illustrations, 15s.

A History of the British Marine Polyzoa. By the Rev. THOMAS HINCKS, B.A., F.R.S. With Plates, giving Figures of the Species and principal Varieties. 2 vols. demy 8vo, £3 3s.

GURNEY & JACKSON, 1, Paternoster Row.

(MR. VAN VOORST'S SUCCESSORS.)

An Invasion of Rooks.

At the last meeting of the British Ornithologists' Club one of the members, Mr. Griffith, remarked that on getting out of the train at Orpington Station, Kent, about 4.20 in the afternoon of November 14 he saw an extraordinary flight of rooks passing in a steady, continuous stream for 16½ minutes, apparently making for Farnborough. They were winging the way in a great column of from fifteen to twenty abreast, and were moving at about ten miles an hour. On the lowest estimate he reckoned that about 13,000 passed over him, but how many had already passed before he stepped out of the railway station he could not, of course, say, but they stretched away southwards as far as the eye could see. Whether these were native birds which had been on a foraging expedition and were returning to their roosting trees or whether this great army were aliens come to take up their quarters here it would be difficult to say. If they were native-bred birds then such a gathering must have some hidden significance, for gatherings on such a grand scale are unprecedented.

Westminster Gazette. Dec. 12. 1910

www.ingramcontent.com/pod-product-compliance
Lightning Source LLC
Chambersburg PA
CBHW030552270326
41927CB00008B/1619